P9-EDX-606

MathStart®
COUNTING COINS

The Penny Pot

by Stuart J. Murphy • illustrated by Lynne Cravath

HarperCollinsPublishers

For Barbara Elleman—who has provided more
than a potful of friendship and support
—S.J.M.

For Leigh and Lauren
—L.C.

The illustrations in this book were created with gouache, ink, and pastels on Arches watercolor paper.

HarperCollins®, ®, and Mathstart® are trademarks of HarperCollins Publishers.

For more information about the MathStart series, please write to
HarperCollins Children's Books, 10 East 53rd Street, New York, NY 10022,
or visit our web site at http://www.harperchildrens.com.

Bugs incorporated in the MathStart series design were painted by Jon Buller.

THE PENNY POT

Library of Congress Cataloging-in-Publication Data
Murphy, Stuart J., date
The penny pot / by Stuart J. Murphy ; illustrated by Lynne Cravath.
p. cm. — (MathStart)
"Level 3, counting coins."
Summary: The face painting booth at the school fair provides plenty of opportunities to count combinations of coins adding up to fifty.
ISBN 0-06-027606-1. — ISBN 0-06-027607-X (lib. bdg.)
ISBN 0-06-446717-1 (pbk.)
1. Counting—Juvenile literature. 2. Coins—Juvenile literature. [1. Counting. 2. Coins.] I. Cravath, Lynne Woodcock, ill. II. Title. III. Series.
QA115.M873 1998 97-19776
513.2—dc21 CIP
AC

1 2 3 4 5 6 7 8 9 10
❖
First Edition

The Penny Pot

It was a hot Saturday in June, and the school fair was very crowded. The busiest place of all was the face-painting booth. Fran, the art teacher, was in charge.

More than anything else, Jessie wanted to get her face painted. It cost 50 cents.

Jessie emptied her pockets and counted her money.

She had three dimes, one nickel, and four pennies.

"Oh," she said sadly. "I have only 39 cents." Now she wished she hadn't bought that ice cream cone.

"Don't worry," said Fran. "People will put their extra pennies in this pot, and you can have them. Wait and see."

So Jessie sat down to wait.

Soon Miguel came along. "I'd like to get my face painted," he said to Fran.

"Sure," she replied. "Do you have 50 cents?"

"Let's see," said Miguel.

He had a quarter, a nickel, two dimes, and three pennies.

"More than enough," said Fran. "Would you like to put your extra pennies into the pot for someone else to use?"

"Sure!" said Miguel.

"Now," said Fran, "what would you like to be?"
"I like clowns," said Miguel. "Can you make me into one?"
In five minutes Miguel looked like the funniest clown around.

Jessie's friend Rachel came by next, with her little sister, Sam. "I want my face painted!" cried Sam.

"Okay, okay," said Rachel. "Let's look in your purse."

There was a quarter, a dime, two nickels, and seven pennies.

"You have enough money!" she told her. "And you even have two pennies left for the penny pot."

"What would you like to be?" Fran asked Sam.

"A star!" said Sam. "Because my mom says I'm her little star."

Fran painted a big purple star on Sam's face. Then she added a twinkly sticker to one point, right on Sam's forehead.

"Now I'm *really* a star!" said Sam.

The next person in line was Jonathan. "I know just how much money I have," he said. "I have 54 cents."

And sure enough, that's what he had:
three dimes, three nickels, and nine pennies.

Jonathan put the four extra pennies into the penny pot.

"And what would you like to be?" Fran asked him.

"Hmm," he said. "I don't really know. Do you have any ideas?"

"I think you'd look good as a bear," she said.

And he did, too.

After a few minutes Annie came along. "I'd like to get my face painted," she said. She put her money on the table and counted it out.

There were two dimes, four nickels, and thirteen pennies.

"Whew," she said. "I have enough, and even some extra pennies to put in the pot."

"I know just what I want to be!" Annie said before Fran could even ask. "I want to be a scary monster."

"I'll do my best," said Fran. When she was done, Annie didn't look like Annie at all.

Just when Jessie was getting very tired of waiting, Fran smiled at her. "Why don't you see what's in the penny pot now?" she said. "Maybe there will be enough, along with your 39 cents."

Jessie took all her money out of her pocket. Then she turned over the penny pot and dumped the money out onto the table.

She counted it once, and then she counted it again to be sure.

"Fifty cents!" she shouted. "Plus one cent left over!"
"You can leave that penny in the penny pot for the next person who needs it," said Fran. "Now, what would you like to be?"

Jessie couldn't make up her mind. A star? A bear? A clown? They were all nice, but she wanted to be something different.

All at once Fran's cat knocked over the jar of blue paint. "Scat, Chester!" said Fran, catching the jar just in time.

And then—just like that—Jessie knew what she wanted to be.

Soon all the kids from school were gathered to look at Jessie's painted face.

"That's the best one ever!" said Annie.

"I think it is," agreed Jessie. "Isn't it, Chester?"

Chester just swished his tail.

FOR ADULTS AND KIDS

If you would like to have more fun with the math concepts presented in *The Penny Pot,* here are a few suggestions:

- Read the story with the child and talk about what is going on in each picture.
- Ask questions throughout the story, such as: "What coins does Jessie have?" "Can you add them up?" "How does Miguel make fifty cents?"
- Place a handful of coins on the table and talk about the value of each. Ask questions such as: "Which coin is a dime?" "How many pennies equal a dime?" "How many nickels?" "Can you combine pennies and nickels to equal a dime?'
- Reread the story together and ask the child to identify the different coins in the story and tell how much each is worth.
- Place a handful of mixed coins on the table and encourage the child to experiment with counting them up in different ways.
- Practice using coins in everyday situations. Help the child choose the coins needed to buy a magazine or a candy bar. Ask the child to help find the correct change needed to ride the bus, buy a stamp, or purchase a gum ball from a machine.

Following are some activities that will help you extend the concepts presented in *The Penny Pot* into a child's everyday life.

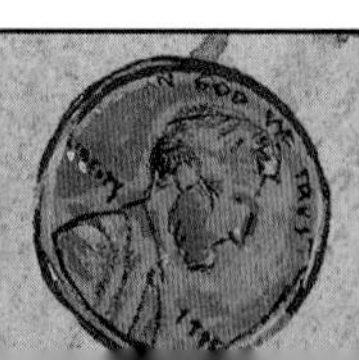

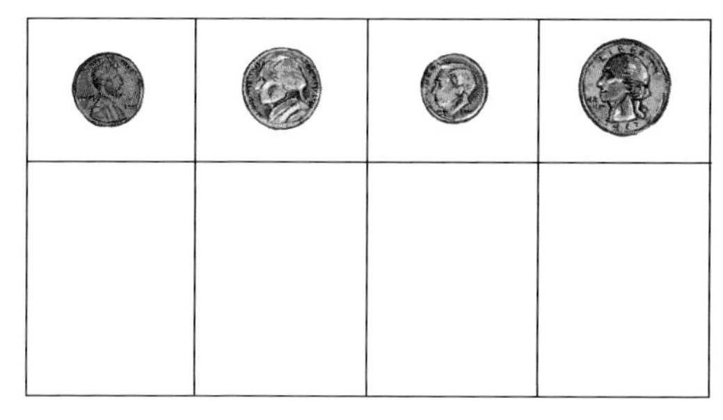

Trading Coins: Construct a mat for each player. Collect one die and a pile of coins with lots of pennies. Each player rolls the die and then receives that number of pennies. Trade five pennies for a nickel. Keep playing and trading up for nickels and dimes. The first player to get a quarter wins.

Make-Believe Shopping: Label kitchen objects with make-believe prices (using the ¢ sign) on stickers or masking tape. Make sure that "customers" are given enough change to make purchases. Make your own cash register with an egg carton to separate the different coins. Now shop!

Playing a Coin Game: Place a handful of coins under a napkin. Tell the other players how many coins you have and what the total is. Then ask the other players to guess what the coins are.

The following stories include concepts similar to those that are presented in *The Penny Pot*:

- BUNNY MONEY by Rosemary Wells
- JELLY BEANS FOR SALE by Bruce McMillan
- ALEXANDER, WHO USED TO BE RICH LAST SUNDAY by Judith Viorst

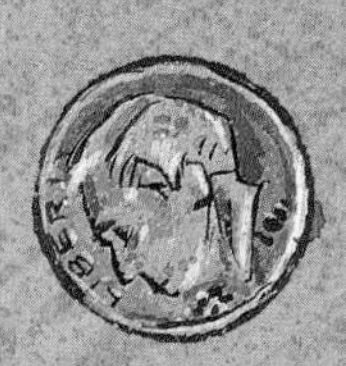

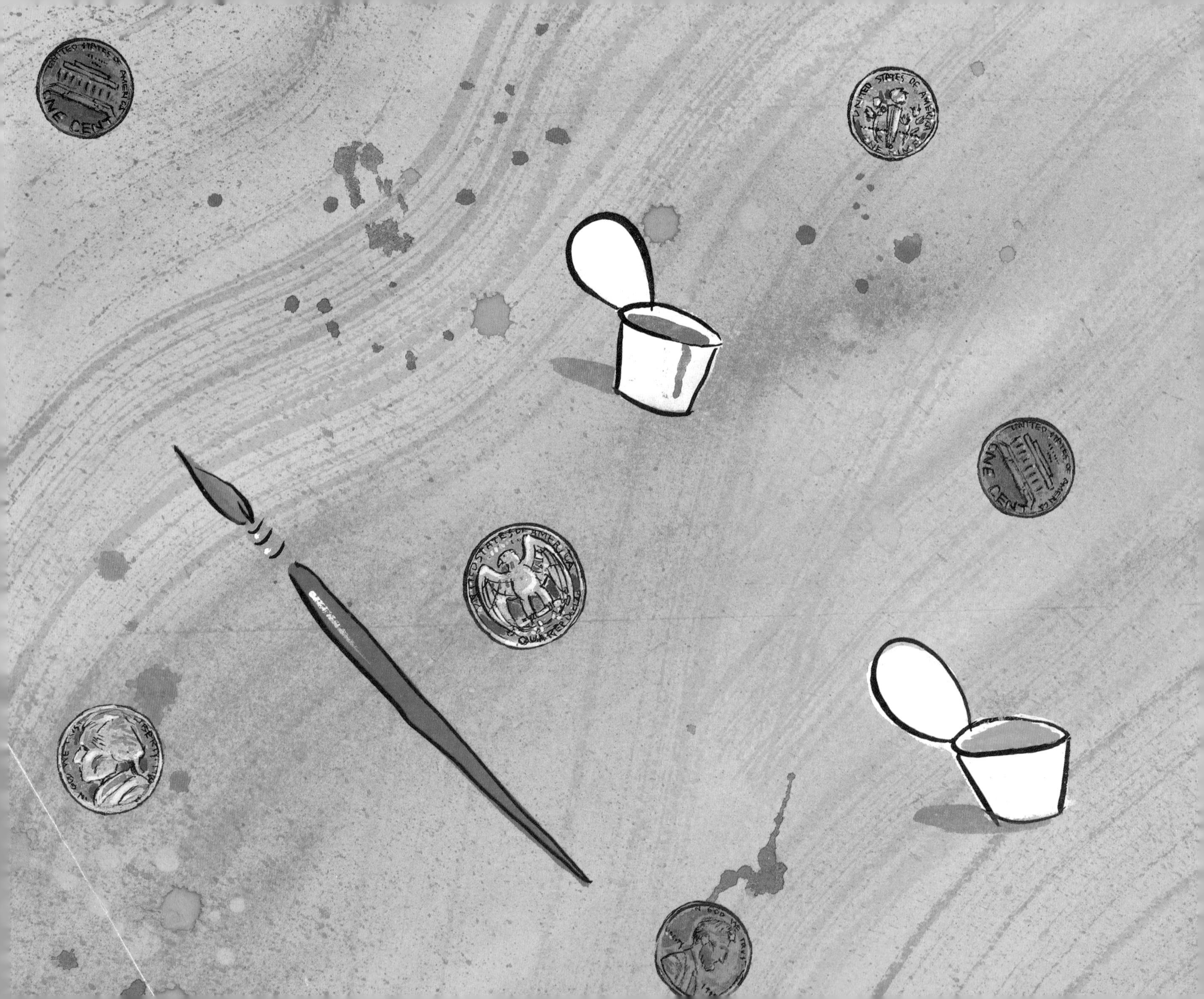
UNITED STATES OF AMERICA
ONE CENT